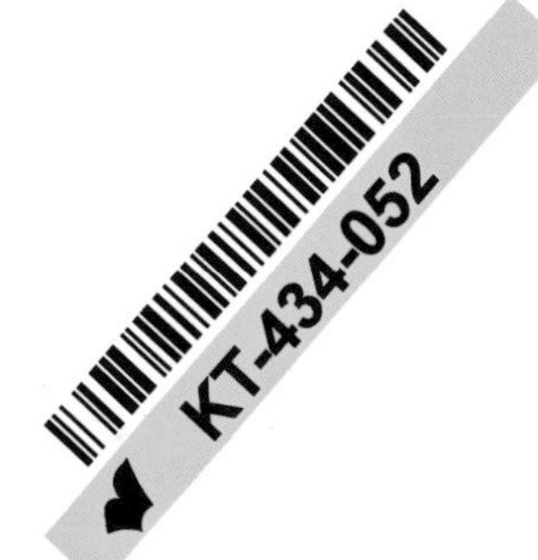

Aspirin

A curriculum resource for post-16 chemistry and science courses

Aspirin (2nd edition)

Compiled by David Lewis

Edited by Colin Osborne and Maria Pack

Designed by Imogen Bertin and Sara Roberts

First published by the Royal Society of Chemistry in 1998
Second edition published by the Royal Society of Chemistry in 2003

Printed by the Royal Society of Chemistry

For further information on other educational activities undertaken by the Royal Society of Chemistry contact:

Education Department
Royal Society of Chemistry
Burlington house
Piccadilly
London W1J 0BA

Email: education@rsc.org
Tel: 020 7440 3344

Information on other Royal Society of Chemistry activities can be found on its websites:
http://www.rsc.org
http://www.chemsoc.org
http://www.chemsoc.org/LearnNet contains resources for teachers and students from around the world.

ISBN 0–85404–388–8

British Library Cataloguing in Publication Data.

A catalogue for this book is available from the British Library.

Contents

How to use this book

This book consists of eight free-standing activities that can be used singly or as a coherent package in a wide range of teaching and learning situations in both academic and vocational courses. It is aimed at post-16 chemistry and science students and their teachers.

The book is organised in two main sections – a series of student activities and a guide for teachers and technicians with equipment lists and answers. Within the section for students there are both written and practical tasks. The practical tasks are both preparative/analytical and problem solving. The student sheets can also be freely downloaded from **http://www.chemsoc.org/networks/learnnet/aspirin.htm**.

Nomenclature

For ease of use some traditional names are retained for chemicals throughout this booklet. Below are listed the traditional names against the systematic names that are commonly used in post-16 chemistry courses in the UK.

Traditional name/trade name	**Systematic name**
Acetanilide	*N*-Phenylethanamide
Aspirin	2-Ethanoyloxybenzenecarboxylic acid
Caffeine	3,7-Dihydro-1,3,7-trimethyl-1H- purine-2,6-dione
Ibuprofen	(±)-2-(4-Isobutylphenyl)propionic acid
Methyl salicylate	Methyl 2-hydroxybenzoate
Paracetamol	*N*-(4-Hydroxyphenyl)ethanamide
Phenacetin	*N*-(4-Ethoxyphenyl)ethanamide
Salicylic acid	2-Hydroxybenzoic acid
Sodium salicylate	Sodium-2-hydroxybenzoate

Health and safety

Teachers must consult their employer's risk assessments before commencing any practical activity. It is good practice to encourage students to do so also. However, this does not absolve teachers from their responsibility to check students' plans and supervise the activity. Eye protection and other appropriate protective equipment should be worn for all the experiments in this booklet. The following texts are available to give guidance on health and safety and assessing risk:

Safeguards in the School Laboratory, 10th edition, ASE, 1996

Topics in Safety, 3rd Edition, ASE, 2001

Hazcards, CLEAPSS, 1995 (or 1998, 2000 updates)

Laboratory Handbook, CLEAPSS, 2001

Safety in Science Education, DfEE, HMSO, 1996

Hazardous Chemicals. A manual for science education, SSERC, 1997 (paper).

Hazardous Chemicals. An interactive manual for science education, SSERC, 2002 (CD-ROM).

Further information can be obtained from:

- CLEAPSS School Science Service at Brunel University, Uxbridge, UB8 3PH; tel 01895 251496; fax 01895 814372; email science@cleapss.org.uk or visit the website **http://www.cleapsss.org.uk** (accessed August 2003).
- Scottish School Equipment Resource Centre (SSERC). Contact SSERC at St Mary's Building, 23 Holyrood Road, Edinburgh, EH8 8AE; tel 0131 558 8180, fax 0131 558 8191, email sts@sserc.org.uk or visit the website **http://www.sserc.org.uk** (accessed August 2003).

Dichloromethane is harmful by inhalation. Avoid breathing vapour and avoid contact with skin and eyes.

Ethanoic anhydride is flammable and causes burns.

Ethanol is flammable.

Ethyl ethanoate is volatile, highly flammable and the vapour may irritate the eyes and respiratory system. Avoid breathing the vapour and avoid contact with the eyes. Keep away from flames.

Hydrochloric acid can cause burns. It gives off an irritating vapour that can damage the eyes and lungs.

2-Hydroxybenzoic acid (salicylic acid) is harmful by ingestion and is irritating to the skin and eyes.

Phenol is toxic by ingestion and skin absorption. It can cause severe burns. Take care when removing phenol from the bottle because the solid crystals can be hard to break up. Wear rubber gloves and a face mask.

Phosphoric acid is irritating to the eyes and causes burns.

Sodium hydroxide can cause severe burns to the skin and is dangerous to the eyes.

Short wave UV may cause skin cancer and eye damage. Do not observe directly. The viewer should be screened from direct radiation.

Background information

1. The aspirin story

Nearly all of us have used aspirin at some time in our lives, but not many of us know that for hundreds of years a related compound from willow bark has been used to relieve pain and treat fevers. Ancient Asian records indicate its use 2400 years ago.

In this activity you are going to find out about the discovery and use of aspirin and to present your findings.

Here is a 'time line' for the past 230 years to get you started.

1763 Edward Stone (a clergyman) read a paper to the Royal Society of London: 'An account of the success of the Bark of the Willow in the Cure of Agues'. He had collected observations from around the country on the effect of willow bark on the relief of fever due to agues (malaria).

1830s A Scottish physician found that extracts of willow bark relieved symptoms of acute rheumatism.

1840s Organic chemists working with willow bark and flowers of the meadowsweet plant, spirea, isolated and identified the active ingredient as salicin (salix = Latin word for willow).

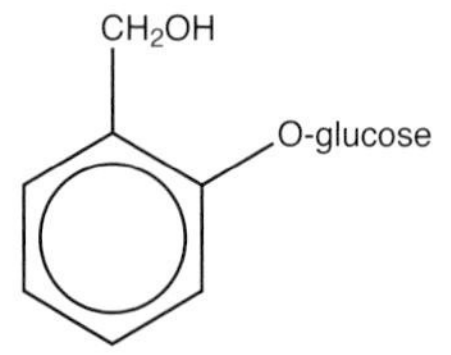

CH_2OH

O-glucose

Salicin

1870 Professor von Nencki of Basle demonstrated that salicin was converted into salicylic acid in the body.

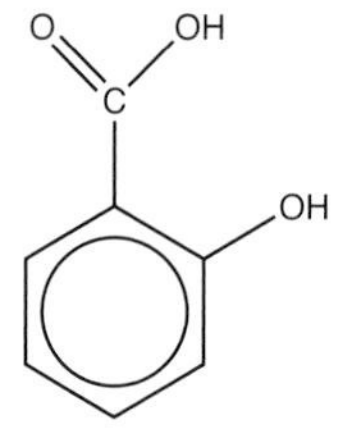

O OH

C

OH

Salicylic acid (2-Hydroxybenzoic acid)

Salicylic acid was then given to patients with fevers and their symptoms were relieved. However, the compound caused severe irritation of the lining of the mouth, gullet and stomach.

1875 Chemists made sodium salicylate and gave that to doctors to try on their patients. It still worked to help reduce pain and fever and did lessen the irritation, but tasted awful!

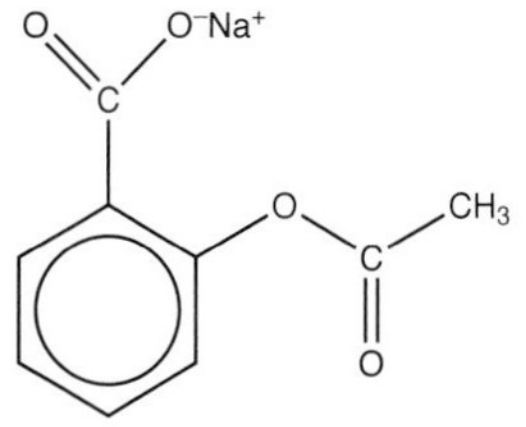

Sodium salicylate
(Sodium 2-hydroxybenzoate)

In the large doses used for treating rheumatism sodium salicylate frequently caused the patient to vomit.

1890s Felix Hofmann of the Bayer Company in Germany made aspirin which was found to have good medicinal properties, low membrane irritation and a reasonable taste. This followed the publication of news about the temperature reducing properties of acetanilide which immediately spurred a chemist at Bayer's dyeworks to make some derivatives:

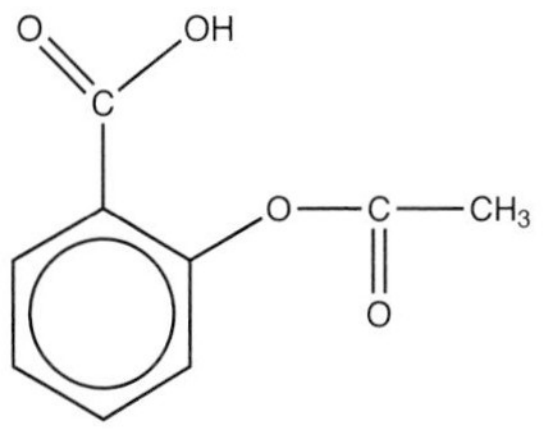

Aspirin
(2-Ethanoyloxybenzenecarboxylic acid)

He called the new medicine aspirin ('a' for acetyl – the systematic name for the compound at the time was acetylsalicylic acid, 'spir' for spirea, the meadowsweet plant).

Nowadays chemists use the systematic name, ethanoyl, instead of acetyl; but the trivial name acetyl is still very common.

1898 Aspirin was sent for *clinical trials*, Bayer manufactured the medicine and patented the process.

1915 During World War I the British wanted aspirin but it was made by the Germans (Bayer & Co). So the British government offered a £20,000 reward to anyone who could develop a workable manufacturing process. This was achieved by George Nicholas, a Melbourne pharmacist, who subsequently gave his tablet the name 'Aspro'.

1982 Nobel Prize for medicine awarded for work on prostaglandins and related compounds. John Vare, who shared the prize with two Swedes, discovered that aspirin and some other painkillers and anti-inflammatory drugs (such as ibuprofen) inhibit a key enzyme in the prostaglandin synthetic pathway.

They therefore stop your body making prostaglandins, some of which stimulate pain receptors and cause inflammation. For more information see **www.nobel.se/medicine/laureates/1982/press.html** (accessed August 2003).

1990s More than 10 million kilograms of aspirin are made in the US each year! Nowadays aspirin is not only used as a painkiller but has also been proposed as effective in reducing the incidence of heart disease.

Activity

Work in small groups to prepare an audio-visual presentation about aspirin. When it is complete you should make your presentation to another group of students.

Alternatively you could prepare and make a poster for the wall of the laboratory or tutor room.

Hints on presentations

In a presentation you should include the following:

- the conditions that aspirin helps to relieve or cure, including technical terms such as analgesic, antipyretic and anti-inflammatory;
- the side effects of aspirin, and the alternative treatments for people who are affected by them;
- how aspirin came to be developed over the past 200 years, including the achievements of those responsible for the main developments;
- the chemistry involved in developing the medicine in a usable form; and
- the nature and importance of clinical trials.

You may find information in reference books, in libraries, in pharmacies and by contacting pharmaceutical companies, or the ABPI (Association of the British Pharmaceutical Industry, 12 Whitehall, London SW1A 2DY, Tel 020 7930 3477 **www.abpi.org.uk**.)

Making a poster

In making a poster the following hints may be useful:

- your poster should be clearly set out, the structure should be clear at a glance;
- people do not like reading a lot of text. Diagrams and flow charts are much easier to take in; text should be readable from at least 2 m;
- explanations should be separate from the main story, perhaps in distinctive boxes; and
- the level must be appropriate for the expected audience: you will need to think about what the audience is likely to know already.

Making an audio-visual presentation

In making an audio-visual presentation the following hints may be useful:

- before you start, make sure you have everything ready and you know how to switch on the OHP or operate the data projector and that it is focussed correctly;
- start the presentation with something designed to capture attention and to help your audience to know what to expect;
- do not read directly from notes: use notes if you need to, but always talk directly to your audience;
- people get bored if they have nothing to do but listen to you talking: make sure that there is always something to look at as well;
- make sure your visual aids are prepared well beforehand: they are a very effective way of getting information across to your audience;
- if you are drawing formulae on a whiteboard or blackboard make sure that you know them by heart (draw them out beforehand): you should not have to keep looking at your notes to make sure that you have got something right;
- remember that you are always more familiar with your subject matter than your audience: give them time to take in what you are saying before going on to the next stage; and
- mannerisms are irritating, so try to stand still, look at your audience and do not wave your hands about, or keep scratching your nose or trip over the OHP lead!

Experimental and investigative section

2. The preparation of 2-hydroxybenzoic acid

Many organic compounds are found in plants. 2-Hydroxybenzoic acid (salicylic acid) can be made from methyl 2-hydroxybenzoate which is obtained as oil of wintergreen by distillation from the leaves of *Gaultheriae procunbers*.

Oil of wintergreen is 98% methyl 2-hydroxybenzoate. This oil can be hydrolysed by boiling with aqueous sodium hydroxide for about 30 minutes. The reaction produces sodium 2-hydroxybenzoate which can be converted into 2-hydroxybenzoic acid by adding hydrochloric acid.

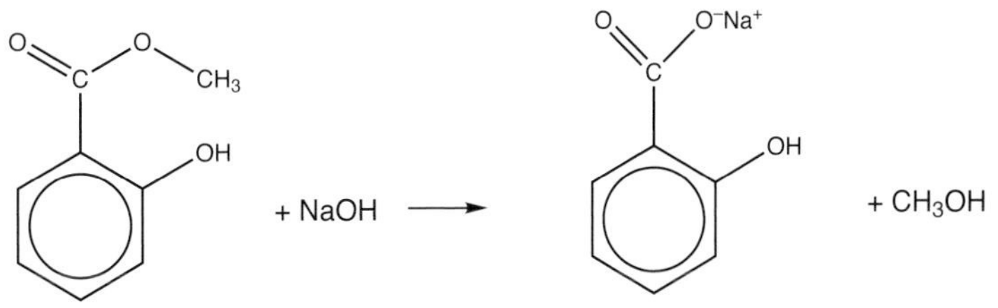

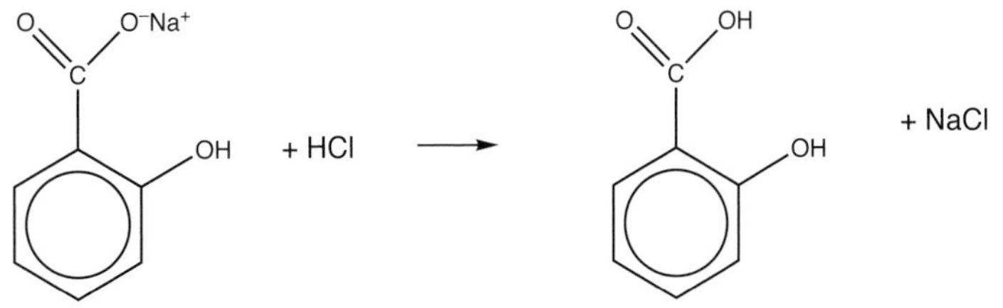

The process has three main stages:

1. heating the oil of wintergreen with aqueous sodium hydroxide. The reaction is quite slow. You need to heat the mixture for 30 minutes without letting the water, or the oil of wintergreen, or the product evaporate. Remember that one of the products, methanol, is flammable so you cannot heat the flask with a naked flame;
2. converting the reaction product into the free acid; and
3. separating the product from the reaction mixture and drying it.

Stage 1

Set up apparatus suitable for heating about 30 cm^3 of reaction mixture using a water bath. Use a condenser to prevent any volatile liquids escaping. Get your apparatus checked by your teacher before you start the reaction. Put on your eye protection.

Put 2 g of oil of wintergreen into your flask and add 25 cm^3 of 2 mol dm^{-3} sodium hydroxide (CARE!). Aqueous sodium hydroxide is particularly prone to bumping so you will need some anti-bumping granules. Then heat over a boiling water bath for 30 minutes.

While the mixture cools make a list of the possible compounds present in the mixture.

Stage 2

Pour the mixture into a small beaker surrounded by a mixture of ice and water. Add concentrated hydrochloric acid (CARE!) to the mixture dropwise until it is just acidic, stirring all the time. Why do you need to keep the mixture cool during this process?

Stage 3

Filter the product using a Buchner funnel and suction apparatus. Wash the product with a little ice cold water and transfer it to a weighed watch glass. Allow to dry overnight.

Results

Include the answers to the following questions in your write up.

1. How can you tell from observing the process that a new substance has been formed in the reaction?
2. What has happened to the methanol formed in the reaction?
3. What is oil of wintergreen used for nowadays?
4. What mass of product was formed from 2 g of oil of wintergreen?
5. What percentage yield is this?

3. The preparation of aspirin

In this activity you use ethanoic anhydride to convert 2-hydroxybenzoic acid into aspirin.

$$C_6H_4(OH)COOH + (CH_3CO)_2O \longrightarrow C_6H_4(OCOCH_3)COOH + CH_3COOH$$

The reaction takes place easily in acidic solution but the product is formed as part of a mixture containing several other compounds. The product is formed in Stage 1 below and then separated from impurities in Stage 2.

Note that ethanoic anhydride reacts readily with water so all the apparatus must be dry.

Stage 1

Take about 1 g of your sample of 2-hydroxybenzoic acid and weigh it accurately. Put it into a dry pear shaped flask and add 2 cm^3 of ethanoic anhydride followed by 8 drops of concentrated phosphoric acid. Put a condenser on the flask.

In a fume cupboard, warm the mixture in a hot water bath, with swirling, until all the solid has dissolved and then warm for another 5 minutes.

Stage 2

Carefully add 5 cm^3 of cold water to the solution. Stand the flask in a bath of iced water until precipitation appears to be complete. It may be necessary to stir vigorously with a glass rod to start the precipitation process.

Filter off the product using a Buchner funnel and suction apparatus.

Wash the product with a little cold water, transfer to a weighed watch glass and leave to dry overnight. Weigh your product.

Results

Include the answers to the following questions in your write up.

1. Calculate the relative molecular masses of 2-hydroxybenzoic acid and aspirin.

   What are the theoretical and actual yields?

   Calculate the percentage yield.
2. How might material be lost at each stage? How could losses be minimised? Why might the apparent yield be too large?
3. What would you expect to be the main impurities in your sample?

Chemical tests for purity

The addition of a chemical reagent to identify substances is a common procedure in chemical analysis. Often the reagent added forms a coloured compound with the substance under investigation. In this experiment iron(III) ions react with one of the possible impurities in the crude aspirin.

Carry out the following tests. Note carefully what you did and all your observations. Then answer the following questions.

- Add 5 cm^3 of water to each of four test-tubes and label them A, B, C and D.
- Dissolve a few crystals of the following substances in water in the test-tubes.
  - **A** Phenol (NB Phenol is a toxic substance; avoid spillage and wash hands after use)
  - **B** 2-Hydroxybenzoic acid
  - **C** Crude product from Activity 3
  - **D** Pure aspirin
- Add 10 drops of a 1% iron(III) chloride solution to each test-tube and note the colour.

Questions

1. Formation of an iron-phenol compound with Fe^{3+} gives a definite colour. Does the crude product contain any phenol type impurities?
2. Draw the structural formulae of phenol, 2-hydroxybenzoic acid and aspirin. Identify the functional group most likely to be reacting with the Fe^{3+} ions.

4. Purifying by recrystallisation

When an organic compound has been made it needs to be purified, particularly if it is a pharmaceutical chemical. This is because most organic reactions produce by-products but, even if the reaction is a 'clean' one, the purity standards for many products are so stringent that small amounts of other compounds have to be removed.

In the laboratory, this is often done by recrystallisation. The general method is to find a solvent that dissolves the product more readily at high temperature than at low temperature, make a hot solution, and allow to crystallise on cooling. The crude product might contain;

- impurities which are insoluble in the solvent;
- impurities which are slightly soluble in the solvent; and
- impurities which dissolve readily in the solvent.

The solvent itself has also to be removed or it behaves as an impurity in its own right. It must not leave behind any residue.

One simple way to tell whether an organic compound is pure is to measure its melting (or boiling) point. A pure compound melts sharply: if impurities are present it melts slowly (over a range of temperature) and the melting point is lower than that of the pure compound.

In this activity, you are going to look at the stages involved in recrystallisation and draw up a procedure for carrying out the process. You may then use your plan to recrystallise a sample of the aspirin that you have made.

On the next page is a list of the stages in recrystallisation, but not necessarily in the correct order. Think about each of the types of impurity above and put these stages in the correct order. Then match each of them to the boxes which describe what has been achieved. You could arrange the results first or you could cut out a copy of the chart and rearrange it.

Stages	Results
Dry the product on a watch glass, either at room temperature or in an oven.	The product dissolves only in the hot solvent. Soluble impurities also dissolve, but there should not be so many impurities that the solution is saturated. Insoluble impurities stay in suspension.
Shake your sample with the solvent and warm to dissolve.	Insoluble impurities stay on the filter paper, soluble impurities and the product stay in solution and are found in the filtrate.
Wash the residue with a small amount of cold solvent. Why do you need to use the solvent cold?	The crystals are separated from the solvent which still contains soluble impurities, leaving the product contaminated only with solvent in which is dissolved a small amount of soluble impurity.
Allow the solution to cool slowly. If no crystals appear, add a single crystal as a 'seed' or stir vigorously for a few minutes. (If still no crystals appear you have probably added too much solvent!)	Because the solvent is pure it leaves no residue apart from the product.
Filter the solution hot. Use a Buchner funnel and pump to make sure that the solution does not cool too much while it is being filtered. Throw away the residue.	Contaminated solvent passes through the funnel, leaving only product and pure solvent on the filter paper. Warm solvent would dissolve a significant amount of product and it would pass through the filter paper.
Filter the solution cold. Use a Buchner funnel and pump. Keep the residue.	The product becomes less soluble as the mixture cools and eventually crystallises. Soluble impurities are less concentrated so they stay in the solution. (NB if the mixture is cooled too quickly solvent can become trapped in the crystals and is difficult to remove.)

When you have done this activity you should try out your method using half of your sample of aspirin. Water is a good solvent to use but you get better results using ethyl ethanoate. You might like to organise your group to compare the two solvents.

Weigh your sample before you start and again afterwards to find out how much you lose in the recrystallisation.

Questions

1. Where do you think most material is lost?
2. Do you think it is mostly product, mostly impurity or some of each?
3. What conclusion would you come to if your sample weighed more after recrystallisation?

In the next activity you measure the melting point of your sample before and after recrystallisation.

5. The melting point of aspirin

Measuring the melting point of a substance is a good way to test for purity. In this experiment you use the melting point as a way of investigating the purity and identity of laboratory prepared aspirin samples.

A pure substance usually has a sharp melting point – *ie* a narrow temperature range during which it changes from a solid to a liquid. A substance which contains impurities often melts over a range of several degrees.

Any impurities in the substance cause a lowering and broadening of this characteristic temperature.

Substance	**Melting point**
2-Hydroxybenzoic acid	158–160 °C
Aspirin	138–140 ° C

Method

1. If you do not have sealed melting point tubes, heat the end of a capillary tube in a Bunsen burner flame until the glass softens and the end is sealed.

   Do not heat the tube so strongly that it bends. Leave it on a heatproof mat to cool.

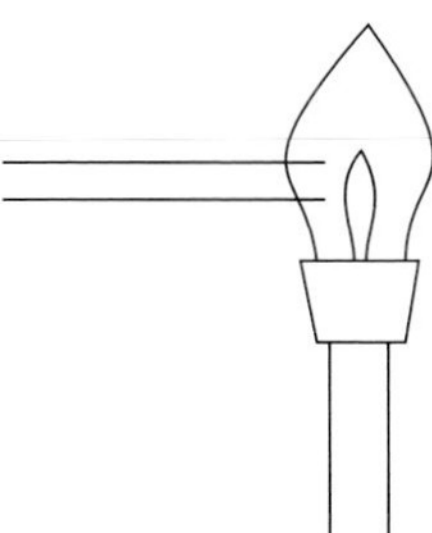

2. Make sure that your samples of solid are dry, by leaving them in a desiccator or an oven at 50 °C overnight. Fill the melting point tube to a depth of about 0.5 cm with dry impure aspirin sample made in Activity 4.
3. Seal a second tube and fill it to a depth of about 0.5 cm with dry crystals of purified aspirin made in Activity 4.
4. Place each tube in the melting point apparatus, slowly increase the temperature and note the temperature range over which the substances melt. A simple but effective apparatus consists of a beaker of oil or glycerol in which is supported a 0–360 °C thermometer. The two melting point tubes are attached to the thermometer close to the bulb using a rubber band. This apparatus makes it easy to compare the behaviour of the two solids.

   An electric melting point apparatus can also be used.

Record the melting point ranges of the pure and impure samples and include a description of the melting process in each case.

Questions

1. On the basis of melting point is it reasonable to conclude that the substances tested contain aspirin?
2. Account for any difference between the melting points of the crude and recrystallised samples of aspirin.
3. What other impurities could there be in the aspirin made in the laboratory?
4. Describe in molecular terms your ideas of what happens when a substance melts.
5. Why does this explanation support the fact that aspirin has a lower melting point than 2-hydroxybenzoic acid?

6. Using thin-layer chromatography to investigate the reaction

You have probably used a simple chromatography experiment as part of your earlier studies to separate the dyes in a coloured ink. The same technique can be used to separate substances which are not dyes but in such experiments the chromatogram must be developed to show up the various different substances that have been separated.

Chromatography techniques are used a great deal in industry because they can be controlled very precisely and use very small amounts of substance. In this activity you investigate the purity and identity of your laboratory prepared aspirin samples using thin-layer chromatography (tlc). In this activity all the substances are white or colourless so you will need to develop the plate before you can see what has happened.

Thin-layer chromatography is a powerful tool for determining if two compounds are identical. A spot of the compound being investigated is placed on a chromatography plate, and a spot of a pure manufactured sample of the same substance is placed next to it. The plate is then allowed to stand in a suitable solvent, which travels up the plate. If the compound to be identified leaves exactly the same pattern on the chromatography plate as the known pure compound it is reasonable to conclude that they are the same. However, if extra spots are observed as well as the characteristic pattern of the known compound, then impurities are likely to be present in the sample.

In this experiment both crude and recrystallised samples of aspirin are compared with a known sample of aspirin.

Method

1. Make sure that you do not touch the surface of the tlc plate with your fingers during this activity. Handle the plate only by the edges and use tweezers if possible.
2. Take a tlc plate and using a pencil (not a biro or felt tip pen) lightly draw a line across the plate about 1 cm from the bottom. Mark three equally spaced points on this line.

3. Place small amounts (about 1/3 of a spatula measure) of your crude aspirin, your recrystallised aspirin and the commercial sample of aspirin in three separate test-tubes. Label the test-tubes so that you know which is which.
4. Make up 5 cm^3 of solvent by mixing equal volumes of ethanol and dichloromethane in a test-tube. Add 1 cm^3 of the solvent to each of the test-tubes to dissolve the samples. If possible do this in a fume cupboard.
5. Use capillary tubes to spot each of your three samples onto the tlc plate. Allow the spots to dry and then repeat three more times. The spots should be about 1–2 mm in diameter.
6. After all the spots are dry, place the tlc plate in the developing tank making sure that the original pencil line is above the level of the developing solvent – ethyl ethanoate. Put a lid on the tank and allow to stand in a fume cupboard until the solvent front has risen to within a few millimetres of the top of the plate.
7. Remove the plate from the tank and quickly mark the position of the solvent front.

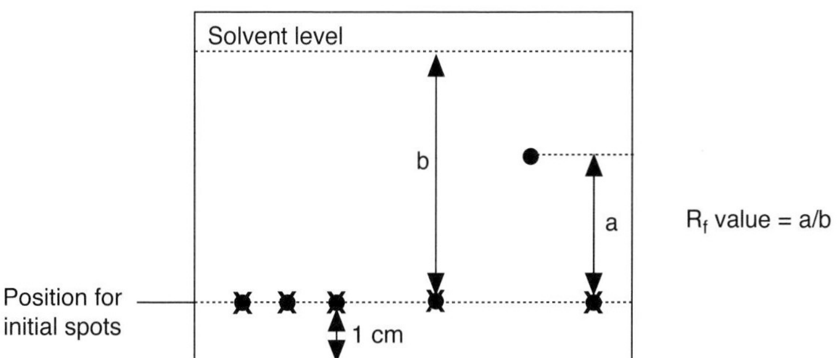

Allow the plate to dry.

8. Observe the plate under a short wavelength UV lamp and lightly mark with a pencil any spots observed.
9. Carefully place the plate in a jar or beaker containing a few iodine crystals. Put a cover on the jar and warm gently on a steam bath until spots begin to appear. Do this in a fume cupboard if possible.

Results

- Draw a diagram to show which spots appeared under UV light and which appear with iodine.
- Determine the R_f value of the samples using the expression R_f = distance moved by sample/distance moved by solvent

Questions

1. Write a short paragraph explaining why some substances move further up the tlc plate than others and how the results are made visible.
2. What conclusions can you draw about the nature of the three samples tested?

7. The solubility of aspirin

The more soluble a medicine is, the more quickly it passes from the digestive system into the bloodstream after being swallowed. In this activity you find the solubility of aspirin by titrating a saturated solution of it with aqueous sodium hydroxide. Aspirin is a weak acid so the solution has a pH greater than 7 at the end-point and phenolphthalein is a suitable indicator to use.

Method

1. Weigh accurately about 0.5 g of aspirin into a 100 cm^3 conical flask, add exactly 50 cm^3 of distilled water (use a pipette or burette) and swirl for 5 minutes. Filter the solution into a dry 100 cm^3 conical flask.
2. Using a 10 cm^3 pipette, transfer four separate 10 cm^3 samples into 100 cm^3 conical flasks.
3. Add four drops of phenolphthalein to each of the flasks and titrate with 0.020 mol dm^{-3} aqueous sodium hydroxide until the first permanent pink colouration. Carry out one rough, and at least two accurate, titrations and record your results in a table.
4. Find the mean volume, V cm^3, of aqueous sodium hydroxide needed to react with the aspirin dissolved in 10 cm^3 of solution.

Calculations

1. Write an equation for the reaction of sodium hydroxide with aspirin. How many moles of sodium hydroxide react with one mole of aspirin?
2. Find the number of moles of sodium hydroxide (NaOH) contained in V cm^3 of 0.020 mol dm^{-3} aqueous sodium hydroxide.
3. Find the number of moles of aspirin contained in 10 cm^3 of solution. What is the mass in grams of this amount of aspirin?
4. How does this compare with the original mass of aspirin?

Conclusions

Write a short paragraph setting out your results. You should explain why it is important for aspirin to be soluble in water and what your results show. How could you attempt to make aspirin more soluble? Carry out a simple experiment to test your suggestion.

Suggest a method you might use to increase the solubility of aspirin without losing its effectiveness as a painkiller. Does the fact that stomach acid contains about 2% hydrochloric acid have any bearing on your conclusions?

8. Thin-layer chromatography

The range of over-the-counter analgesics (painkillers) is quite extensive. In this activity you use thin-layer chromatography (tlc) to separate the components of these medicines and, by comparing the samples, positively identify them.

In the first part of the activity you will make a reference plate. Then samples of analgesics are run on a separate plate and compared with the reference.

Method

You need two tlc plates and six capillary tubes as micropipettes. Make sure that you do not touch the surface of the tlc plates with your fingers during this activity. Handle the plate only by the edges and use tweezers if possible.

Part A

1. Take a tlc plate and using a pencil (not a biro or felt tip pen) lightly draw a line across the plate about 1 cm from the bottom. Mark off three equally spaced points.
2. You are provided with reference solutions which contain, respectively, aspirin, caffeine and a known mixture of aspirin and caffeine. Use three of the micropipettes to spot samples of these reference solutions onto the tlc plate. Allow the spots to dry and then repeat three more times. The spots should be about 1–2 mm in diameter.

O
CH_3
H_3C
N
N
O
N
N
CH_3

Caffeine

3. When all the spots are dry, place the tlc plate in the developing tank making sure that the original pencil line is above the level of the developing solvent – ethyl ethanoate. Put a lid on the tank and allow to stand in a fume cupboard until the solvent front has risen to within a few millimetres of the top of the plate.
4. Remove the plate from the tank and quickly mark the position of the solvent front. Allow the plate to dry.
5. Observe the plate under a short wavelength UV lamp and lightly mark with a pencil any spots observed.
6. Place the plate in a jar or beaker containing a few iodine crystals. Put a cover on the jar and warm gently on a steam bath until spots begin to appear. Do this in a fume cupboard.

Part B

1. Prepare a tlc plate with four points on the base line.
2. Place half a tablet of one of the analgesics to be analysed on a piece of paper and crush it with a spatula. Transfer it to a small labelled test-tube and add 5 cm^3 of the solvent (a 1:1 mixture of ethanol and dichloromethane). Warm gently on a steam bath to dissolve as much of the tablet as possible. Any residue is likely to be a binding agent: allow it to settle for a few minutes. This may be starch. How could you confirm this? Repeat this procedure to make solutions of the other two analgesics.
3. Using similar procedures to Part A 2, spot a sample of each of the clear solutions onto your prepared tlc plate. On the fourth spot place the reference mixture used in Part A.
4. Repeat 3–6 from Part A. Draw diagrams to show how your results appear under UV light and in iodine.

Conclusions

Set out your results to show which of the tablets contain, aspirin, which contain caffeine and which, if any, contain other compounds. What is the purpose of the caffeine?

Discuss some of the advantages and disadvantages of using tlc in the analysis of medicines.

Teachers' and technicians' guide to the activities

1. The aspirin story

The history of aspirin and other medicines dealing with pain, fever or inflammation reveals many interesting points about scientific methodology and the interaction of people and society with technology. One overriding theme that emerges when looking at the development of medicines is the importance of sharing information and cooperation in research. Time and again discoveries in one part of the world have been published, but not developed fully until another person reads and uses the information in another time and place.

The emphasis in this activity is on students finding out for themselves and presenting their findings to a suitable audience. The teacher's role is to be supportive and to offer advice.

The students' worksheet gives some detail of the history of the development of aspirin and related medicines, but it is intended that students find out more than is given to them here. For those who do not have access to relevant reference material or who need additional help, some additional information is set out below.

Conditions that aspirin helps to cure

- Pain – analgesic
- Fever – antipyretic
- Inflammation – anti-inflammatory; and
- Rheumatism – antirheumatic.

Side effects

- Aspirin can lead to stomach ulcers
- Alternative pain relief is available: see below.

Chemistry of aspirin

The presentation should give the basic structures of salicin, salicylic acid and its derivatives. The equation showing that aspirin reacts as an acid with aqueous sodium hydroxide should be given.

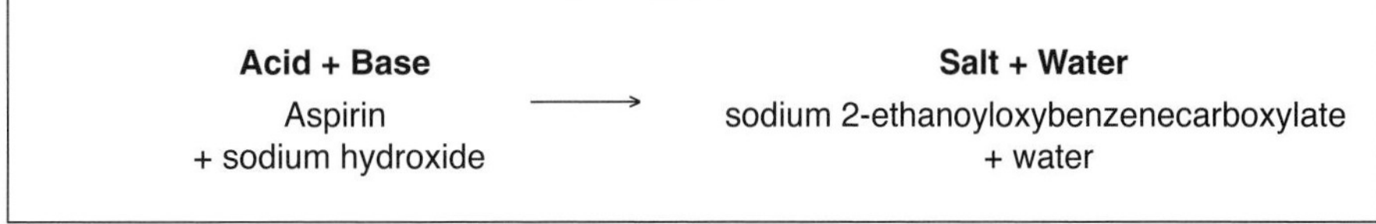

Acid + Base		**Salt + Water**
Aspirin + sodium hydroxide	⟶	sodium 2-ethanoyloxybenzenecarboxylate + water

Nomenclature

- A variety of names are commonly used. Generally the everyday or industrial names are less complex than the systematic names.

  In this publication the name aspirin is retained. The systematic name listed on page 2 is in accord with the ASE document Signs, Symbols and Systematics (2000) and aspirin is thus called

  2-ethanoyloxybenzenecarboxylic acid, rather than

  2-acetoxybenzoic acid or acetylsalicylic acid.

Methods of establishing the safety and efficacy of medicines

- 'Clinical trial' – usually near the end of the medicine testing procedure where the compound is tried out on one group of patients and compared with the effect of a placebo on another group.
- In the late 1800s compounds were given to patients almost immediately after synthesis or discovery.
- A common way of testing anti-inflammatory action is to irritate the joint of a rat's leg until it is inflamed and then administer the medicine.
- Students could be asked to consider the ethical considerations of animal testing.

Alternative pain relief

Paracetamol

At the University of Straßburg in the 1880s Professor Kußmaul, of the Department of Internal Medicine, asked two assistants to give naphthalene as a treatment for intestinal worms.

The medicine had little effect on worms, but one patient had a great reduction in fever temperature. It was found that this patient had, in fact, been given acetanilide instead of naphthalene due to a mistake at the pharmacy!

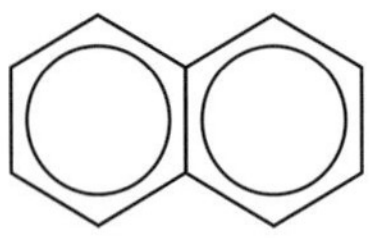

O
HN—C—CH_3

Naphthalene

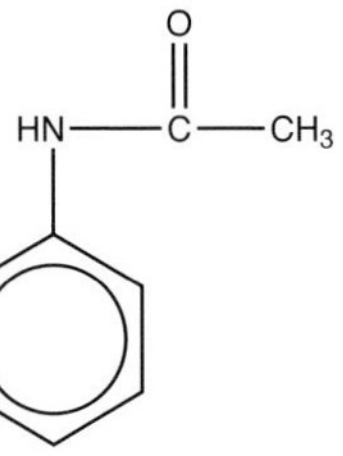

Acetanilide (*N*-Phenyl-ethanamide)

The young assistants quickly published the discovery of this new antipyretic. It was soon in production and remained in use for several years because it was so cheap to produce. However, it had a serious side effect involving the deactivation of some of the hæmoglobin in red blood cells.

The publication of news about acetanilide immediately spurred a chemist at Bayer's dyeworks to make some derivatives:

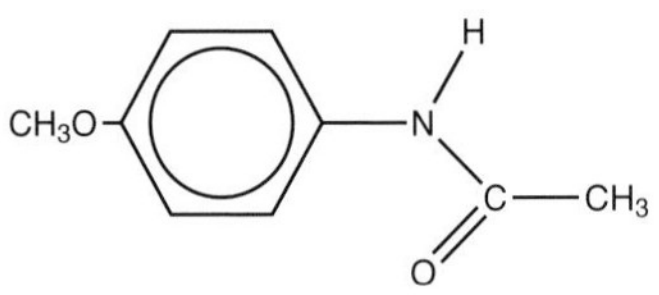

CH_3O– H N C—CH_3 O

CH_3CH_2O– H N C—CH_3 O

N-(4-Methoxyphenyl)ethanamide

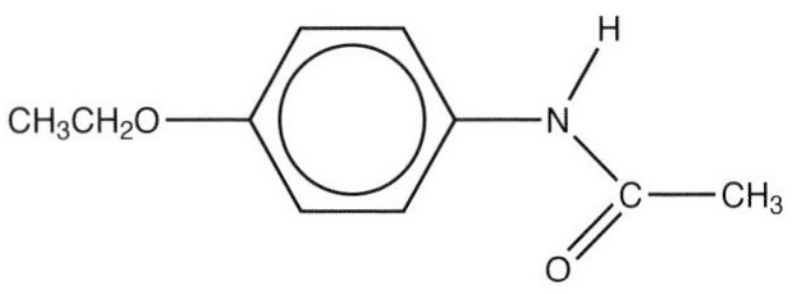

N-(4-Ethoxyphenyl)ethanamide
(Phenacetin)

These were both found to be antipyretic and *N*-(4-ethoxyphenyl)ethanamide was less toxic than acetanilide itself. It was promptly marketed as 'Phenacetin' and has remained in use ever since. However, restrictions have been placed on its use due to kidney damage in long-term users.

Many medicines were synthesised to try to improve on phenacetin and as early as 1893 Joseph von Mering made paracetamol.

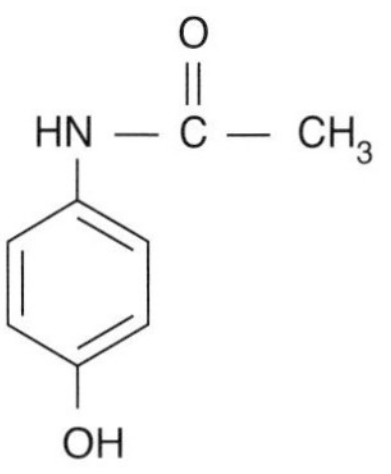

N-(4-Hydroxyphenyl)ethanamide
(Paracetamol)

He found it to be an effective antipyretic and analgesic, but wrongly thought that it caused the same hæmoglobin problem as acetanilide.

It was not until the 1940s that paracetamol was reinvestigated after it was found present in patients dosed with phenacetin. In 1953 paracetamol was marketed by Sterling-Winthrop Co., and promoted as preferable to aspirin since it was safe to take for children and people with ulcers. However, it causes liver damage from chronic use.

Ibuprofen

In the 1960s researchers at Boots decided to synthesise a series of compounds with the aim of producing an alternative to aspirin. They based their new compounds around the benzene ring and carboxylic acid group of aspirin. More than 600 compounds were made and tested before ibuprofen was chosen as the medicine to market.

In 1983, after 15 years of use, ibuprofen became available as an over-the-counter medicine (not just on prescription) due to its minimal side effects.

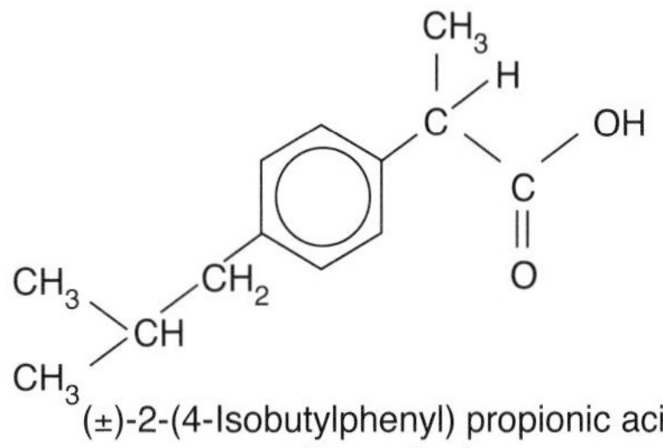

(±)-2-(4-Isobutylphenyl) propionic acid
(Ibuprofen)

Ibuprofen – a case study in green chemistry

The original Boots synthesis has now been superseded by a more environmentally friendly (or green) synthesis. Full details can be found at **http://www.chemsoc.org/networks/learnnet/green/index2.htm**.

2. The preparation of 2-hydroxybenzoic acid

Apparatus

- One 10 cm^3 measuring cylinder
- One 50 cm^3 pear shaped flask fitted with a reflux condenser
- Anti-bumping granules
- Water bath

NB the reaction can be heated directly over a low Bunsen burner flame but care must be taken not to heat too strongly, causing bumping

- One 100 cm^3 beaker surrounded with ice and water in a larger beaker, stirring rod
- Buchner flask and suction apparatus
- Watch glass.

Chemicals

- Oil of wintergreen
- 50 cm^3 of 2 mol dm^{-3} aqueous sodium hydroxide
- Red litmus paper.

Answers to questions

1. There are two layers in the flask at the start of the preparation but when the reaction is complete the mixture is homogeneous.
2. The methanol produced remains in the filtrate because it is completely miscible with water.
3. Oil of wintergreen is 98% methyl salicylate and it shows the medicinal properties of salicylates in general. It is not usually given by mouth but is readily absorbed by the skin. It is applied as a liniment for the relief of pain in lumbago, sciatica and rheumatic conditions, as well as for minor sports injuries.
4. A good yield can be obtained in this experiment. Two grams produces 1.7 g of dry 2-hydroxybenzoic acid.

Further investigations

- The melting point of 2-hydroxybenzoic acid could be determined (158–160 °C)
- Other examples of ester hydrolysis could be investigated – *eg* soap making
- Other natural oils could be extracted from natural materials using steam distillation – *eg* limonene from orange peel.

3. The preparation of aspirin

Apparatus

- Access to a fume cupboard
- One 25 cm^3 pear shaped flask
- Hot water bath
- One 10 cm^3 measuring cylinder
- Bath of iced water
- Glass stirring rod
- Buchner funnel and suction apparatus
- Watch glass.

Chemicals

- 1 g of 2-Hydroxybenzoic acid
- 2 cm^3 of Ethanoic anhydride
- Eight drops of concentrated phosphoric acid.

Notes

Sulfuric acid can be used in place of phosphoric acid but may give lower yields.

Some teachers have reported problems which were due to using ethanoic anhydride that had already been hydrolised to ethanoic acid. Add a drop to water to ensure it is still reactive.

If no precipitate appears, scratch the inside of the beaker with a glass rod or add a seed crystal of aspirin.

As much as 40% of the mass of product after filtering may be water. Overnight drying is preferable to oven drying.

Students should obtain about 0.9 g of crude product from 1.0 g of 2-hydroxybenzoic acid.

Relative molecular masses are:

2-hydroxybenzoic acid:	138
ethanoic anhydride:	102
aspirin:	180

Further investigations

Vary the reaction conditions to investigate the effect on percentage yield of:

- type of acid catalyst;
- concentration or volume of acid used;
- time of heating/cooling; and
- relative amounts of reagents.

Use thin-layer chromatography (tlc) to investigate the purity of the product, using commercial aspirin as a reference. Ensure that this aspirin sample is not 'soluble aspirin' (sodium or calcium salt).

Tests for impurities

- Three test-tubes together with means of labelling them;
- A few crystals of the following substances:
  - **A** Phenol (NB phenol is a toxic substance; avoid spillage and wash hands after use)

B 2-Hydroxybenzoic acid (salicylic acid)
C Crude product from Activity 3
D Pure aspirin

- One per cent iron(III) chloride solution.

Results

Phenol + Fe^{3+}(aq)	Purple solution
2-Hydroxybenzoic acid + Fe^{3+}(aq)	Purple solution
Crude product + Fe^{3+}(aq)	May have a purple tinge due to unreacted 2-hydroxybenzoic acid
Pure product + Fe^{3+}(aq)	Very pale yellow

Answers

1. The crude product may contain 2-hydroxybenzoic acid, as well as water or ethanoic acid as impurities. 2-Hydroxybenzoic acid can be formed either from incomplete reaction or from hydrolysis of the product during its isolation.

2.

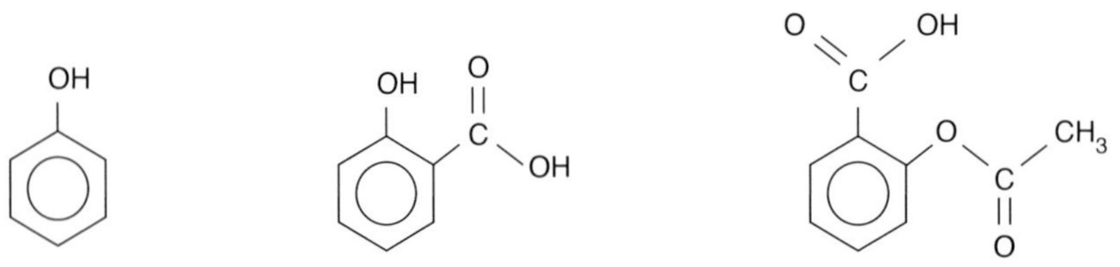

Phenol

Salicylic acid
(2-Hydroxybenzoic acid)

Aspirin
(2-ethanoyloxybenzenecarboxylic acid)

The OH group attached to the benzene ring produces a purple colour with Fe^{3+}(aq) ions. The OH group in aspirin is part of the carboxylic acid group and does not react in the same way.

4. Purifying by recrystallisation

Correct responses

Stages	Results
Shake your sample with the solvent and warm to dissolve.	The product dissolves only in the hot solvent. Soluble impurities also dissolve, but there should not be so many impurities that the solution is saturated. Insoluble impurities stay in suspension.
Filter the solution hot. Use a Buchner funnel and pump to make sure that the solution does not cool too much while it is being filtered. Throw away the residue.	Insoluble impurities stay on the filter paper, soluble impurities and the product stay in solution and are found in the filtrate.
Allow the solution to cool slowly. If no crystals appear, add a single crystal as a 'seed' or stir vigorously for a few minutes. (If still no crystals appear you have probably added too much solvent!)	The product becomes less soluble as the mixture cools and eventually crystallises. Soluble impurities are less concentrated so they stay in the solution. (NB if the mixture is cooled too quickly solvent can become trapped in the crystals and is difficult to remove.)
Filter the solution cold. Use a Buchner funnel and pump. Keep the residue.	The crystals are separated from the solvent which still contains soluble impurities, leaving the product contaminated only with solvent in which is dissolved a small amount of soluble impurity.
Wash the residue with a small amount of cold solvent. Why do you need to use the solvent cold?	Contaminated solvent passes through the funnel, leaving only product and pure solvent on the filter paper. Warm solvent would dissolve a significant amount of product and it would pass through the filter paper.
Dry the product on a watch glass, either at room temperature or in an oven.	Because the solvent is pure it leaves no residue apart from the product.

Introduction

Selecting a good solvent is the key to performing a successful recrystallisation. Ideally, the material to be recrystallised should be sparingly soluble at room temperature and yet quite soluble at its boiling point. Water can be used for recrystallising aspirin because it is cheap, readily available and safe. However, heating aspirin in water partially decomposes it although the quantity of crystals obtained may be satisfactory. The product can be tested with Fe^{3+}(aq) to see whether any improvement in purity can be detected.

Conclusions

- Aspirin crystallises as fine white or transparent needles.
- Some of the desired material is always lost along with the impurities in a crystallisation. The technique can work only if the impurities in the crude product make up a small fraction of the total mass or have very different solubilities.
- A pure substance forms crystals with a characteristic shape. Impurities disrupt the regular array of molecules in a crystal and so lead to irregularly shaped crystals.

Further investigations

- Try further recrystallisation of the product to see if even better quality crystals are produced.
- If water was used as the solvent try recrystallising from ethyl ethanoate.
- Solubility depends on the polarity of the solvent and solute. Investigate the solubility of aspirin in different solvents, predicting the solubility based on the structure of the solvent molecules.

Answers

1. Most material is lost in the solvent.
2. It is mostly impurity that is lost.
3. The sample was contaminated with solvent.

5. The melting point of aspirin

Introduction

A pure substance usually has a specific melting point – *ie* a narrow temperature range during which it changes from a solid to a liquid.

Any impurities in the substance cause a lowering and broadening of this characteristic temperature.

Substance	Melting point
2-Hydroxybenzoic acid	158–160 °C
Aspirin	138–140 ° C

Apparatus

- Melting point tubes
- Watch glass
- Bunsen burner and heatproof mat
- Melting point apparatus: this may consist of a small beaker containing oil or glycerol in which is supported a 0–360 °C thermometer with small rubber bands for attaching the melting pointing tubes, or a commercial apparatus.

Answers

1. This answer will depend on the students' results.
2. The crude product may have a significantly lower melting point (126–132 °C) than pure aspirin as it contains certain impurities.
3. Possible impurities are water, ethanoic anhydride, ethanoic acid, 2-hydroxybenzoic acid and polymerised 2-hydroxybenzoic acid.
4. Forces within crystals of 2-hydroxybenzoic acid may include hydrogen bonding, dipole-dipole forces between polar molecules, and van der Waals forces. These are listed in order of decreasing strength. These forces hold the molecules together in a crystal lattice. When a substance melts these forces have to be overcome. By heating the solid the molecules are given sufficient kinetic energy to overcome these intramolecular forces.
5. The two molecules are the same apart from one having a phenolic OH group and the other having an ester group.

$$—O—\overset{\overset{\displaystyle O}{\|}}{C}—CH_3$$

The phenol can form hydrogen bonds with other molecules whereas the ester cannot. This means that more energy is required by the phenol to overcome these intramolecular forces.

Further investigations

An extension to this Activity might include determining mixed melting points, using 2-hydroxybenzoic acid, the crude aspirin from activity 3 and the purified aspirin.

6. Using thin-layer chromotography to investigate the reaction

Introduction

Revision of chromatography ideas introduced pre-16 could be useful – *eg* separation of coloured inks using filter paper. Emphasise the ability of chromatograms to identify substances from very small samples.

Thin-layer chromatography is a powerful tool for determining if two compounds are identical. If the compound to be identified leaves exactly the same pattern on a chromatography plate as a known compound it is reasonable to conclude that they are the same. However, if extra spots are observed as well as the characteristic pattern of the known compound, then impurities are likely to be present in the sample. The definitive test to show that two pure samples are the same is to run a mixed spot using a variety of solvents and show that you do not get a separation.

In this experiment both crude and recrystallised samples of aspirin are compared with a known sample of aspirin.

Apparatus

- Thin-layer chromatography plate and a pencil (not a biro or felt tip pen)
- Four test-tubes in a stand: method of labelling the test-tubes
- Three capillary tubes for use as micropipettes
- Chromatography chamber. Either a screw top jar tall enough to take the tlc plate, a small beaker with a petri dish for a lid, or a commercial tank
- Access to a fume cupboard and short wavelength UV lamp.

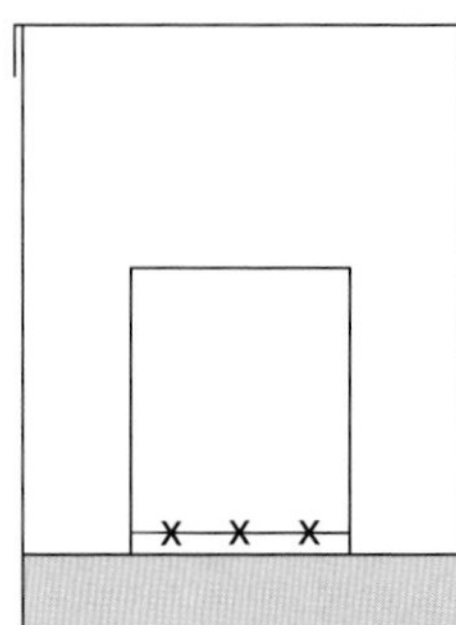

Chemicals

- Ethanol and dichloromethane
- Access to a few iodine crystals (three or four per experiment is enough)
- Samples for testing
- Ethyl ethanoate as chromatography solvent.

Results

A good separation is obtained which can be seen by using UV light to observe the plate. Impurities should be clearly visible in the crude sample.

Answers

1. Separation of substances is based on the many equilibrations the solutes experience between the moving (ethyl ethanoate) and stationary (silica) phase. Less polar substances move more quickly than more polar ones. In general the stationary phase is strongly polar and strongly binds polar substances. The moving liquid phase is usually less polar than the adsorbent and most easily dissolves substances that are less, or even non-polar. The results are made visible by UV absorption or by chemical reaction with iodine.

2. The recrystallised aspirin and the commercial sample should only show one spot with the same R_f value. Other spots should be seen in the crude sample.

Further investigation

Students could:

- try using different solvents to see whether better separation of impurities in the crude product is possible.

7. The solubility of aspirin

Introduction

This activity is a useful vehicle for reinforcing ideas relating to strong and weak acids, basicity and elementary chemical calculations. The worksheet leads students through the calculations, but it may be necessary to explain that not all of a sample of commercial aspirin is likely to dissolve. Students should also understand why phenolphthalein indicator is used in this experiment. A visual representation of this is available on the *RSC Database* CD-ROM.

Apparatus

- Six 100 cm^3 conical flasks (at least one of them must be dry)
- Filtration apparatus
- White tile
- Burette and stand
- 25 cm^3 (or 50 cm^3) and 10 cm^3 pipettes
- Access to a balance reading to 0.01 g.

Chemicals

- Phenolphthalein indicator
- 0.020 mol dm^{-3} aqueous sodium hydroxide.

Results

- Using 0.020 mol dm^{-3} sodium hydroxide and 10 cm^3 samples of dissolved aspirin the titres should be about 9 cm^3 of sodium hydroxide at room temperature based on the solubility of aspirin being 0.33g in 100 cm^3 at room temperature.
- The relative molecular mass of aspirin is 180 g mol^{-1}
- The solubility varies significantly with temperature and is in the range 0.2–0.4 g/100 cm^3 at room temperature.

Conclusions

Increased solubility can be obtained by making the sodium or calcium salt by adding sodium hydroxide or calcium hydroxide to the mixture, or by raising the temperature.

Sodium acetylsalicylate or calcium acetlysalicylate are the usual forms of soluble aspirin sold. They reform aspirin when they come into contact with stomach acid and crystals of aspirin may then irritate the stomach lining.

Further investigations

- Repeat the activity, but leave the mixture to stand overnight to see whether more of the aspirin dissolves. Hydrolysis of the aspirin is likely to take place. How will this affect the results obtained? Try testing for 2-hydroxybenzoic acid in the solution.
- Plan and perform an experiment to find the solubility of aspirin at body temperature (37 °C).
- Find out about the problems of stomach ulcers caused by aspirin.

8. Thin-layer chromatography

Introduction

It is useful to introduce this activity by discussing the use of painkillers and their other effects: antipyretic, anti-inflammatory and antirheumatic. Establish which brands are most commonly used and use these in the analysis.

Apparatus

- Thin-layer chromatography plate and a pencil (not a biro or felt tip pen)
- Test-tubes in a stand: method of labelling the test-tubes
- Capillary tubes for use as micropipettes
- Chromatography chamber. Either a screw top jar tall enough to take the tlc plate, a small beaker with a petri dish for a lid or a commercial tank
- Access to a fume cupboard and short wavelength UV lamp.

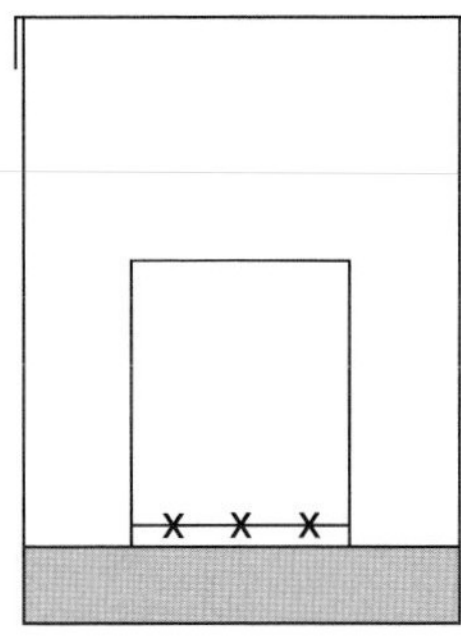

Chemicals

- Dissolving solvent: ethanol and dichloromethane
- Aspirin standard: 1 g aspirin in 20 cm^3 dissolving solvent
- Caffeine standard: 1 g of caffeine in 20 cm^3 dissolving solvent
- Reference standard: a 1:1 mixture of the aspirin and caffeine standards
- Ethyl ethanoate as chromatography medium.

Conclusion

- Analgesics contain aspirin and should produce a spot corresponding to the known aspirin standard.
- Analgesics containing caffeine should produce a spot corresponding to the known caffeine reference.
- Any spots not corresponding to aspirin or caffeine represent other medicines such as ibuprofen. This is less polar than aspirin and therefore moves further up the tlc plate.

(±)-2-(4-Isobutylphenyl) propionic acid
(Ibuprofen)

- Caffeine is a central nervous system stimulant and, in low doses, induces wakefulness and improves mental sharpness.

- Advantages of tlc in medicine analysis:
  - able to separate closely related compounds;
  - only small samples required;
  - quick and easy to carry out; and
  - cheap.

Further investigations

- Give the students an unknown sample to identify.
- Do starch tests on tablets to verify the nature of the binder. Why is starch used?